One-Handed Fractions

By: Karen D. Tollefson

AuthorHouse™
1663 Liberty Drive
Bloomington, IN 47403
www.authorhouse.com
Phone: 1-800-839-8640

Published by AuthorHouse 9/20/2012

ISBN: 978-1-4772-6433-1 (sc)
ISBN: 978-1-4772-6432-4 (e)

Library of Congress Control Number: 2012915709

One-Handed Fractions is a short book to help people who are going back to school, want to get a GED, or just need or want to know how to do the operations done with fractions almost on a daily basis.

This book is intended for any student who for whatever reason did not learn fractions earlier in their lives. The addition and subtraction of fractions may look different than you remember. Trust me when I say this method is easier!!!

I call this One-Handed Fractions because all the operations can be done with 5 steps. Addition and Subtraction are done with the same 5 steps. Multiplication has its own 5 steps as does division.

Move through this book as fast or slow as you want. Some of these skills will come back to you and that will allow you to move faster. Do not get frustrated with any one skill. Get someone to help you or go on to another skill and come back.

You no longer have to skip the fraction problems or be afraid to try them. Good luck to all and I hope this program helps you to understand fractions better.

Karen Tollefson, MEd
Math Teacher

Table of Contents

Content **Page #'s**

1. Why I put this book together 3
2. Fraction Vocabulary 5 – 8

Operations

3. Addition of Fractions 10 – 14
   Addition of Whole and Mixed Numbers 30 - 35
4. Subtracting Fractions 19 -23
   Subtraction of Whole and Mixed Numbers 37 - 40
5. The Rules Are a Changing! 45
6. Multiplication of Fractions 46 – 52
   Multiplication of Whole and Mixed Numbers 59 – 62
7. The Rules Are a Changing! 67
8. Division of Fractions 68 – 72
   Division of Whole and Mixed Numbers 79 – 82

Skill Stops

9. Help to reduce fractions 15 – 17
10. Change Whole or Mixed number to improper fraction 28, 57, 78

Helpful Hints

11. Addition and subtraction of fractions 24, 43
12. Addition, subtraction, and multiplication rules 56
13. Multiplication of fractions rules 65
14. Division of fractions rules 73
15. Rule reminders for mixed numbers 76

Mixed operations practice problems with answers

16. Add and subtract 25 – 27
17. Add and subtract with Whole and Mixed Numbers 41 – 42
18. Just fractions with add, subtract, and multiply 53 – 55
19. Add, subtract, and multiply with Whole and Mixed #'s 63 – 64
20. Just fractions with add, subtract, multiply, and divide 74 – 75
21. Add, subtract, multiply, and divide with Mixed Numbers 83 – 84

Final Review 85 - 86

Vocabulary

These words are not arranged in ABC order, they are arranged in the order you will come to them.

When you see this symbol: *** it means that there is an important piece of information.

- There are also **SKILL STOPS** included.

1. **Fraction**-part of a whole

***The whole can be one piece as a piece of candy, or it can be a group as a group of people.

2. **Denominator**- at the bottom of the fraction: $\frac{}{6}$

- tells how many pieces the whole is cut into

3. **Numerator**- at the top of the fraction: $\frac{3}{}$

- tells the number of the pieces you have

4. **Proper fraction**-the numerator of the fraction is smaller than the denominator. $\frac{3}{5}$

5. **Reduced fraction**-a fraction whose numerator and denominator have no common factors. $\frac{10}{21}$

6. **Factor**-a number that divides evenly into another number.

   Example: 27 1, 3, 9, and 27 are factors of 27 because each divides evenly into 27.

***Be careful here that you do not over reduce. The numerator and denominator can have factors, but they must be in common to reduce. Do not divide them by different numbers.

*** Your answer to a problem involving fractions should always be left as a reduced, proper fraction unless you have another use for the answer such as plugging the answer into a bigger equation.

7. **Improper fraction**-a fraction whose numerator is larger than its denominator. $\frac{21}{4}$

***In higher mathematics, you almost always leave an improper fraction as your answer, but it must be reduced. For example, you would not leave $\frac{25}{15}$ as your answer, you would reduce to $\frac{5}{3}$.

8. **Whole number**-no parts 27

9. **Mixed number**-a whole number and a fraction. $13\frac{1}{4}$

SKILL STOP

- **Change a whole number to a fraction**;----Put the whole number over 1.

  $16 = \frac{16}{1}$ $4 = \frac{4}{1}$ $51 = \frac{51}{1}$

- **Change a mixed number to an improper fraction**:----Multiply the whole number times the denominator, and then add the numerator.

  $4\frac{3}{4} = \frac{19}{4}$ whole # times the denominator 4*4=16 + the numerator 3 = 19

  The new denominator is the same as the old one.

- **Change an improper fraction to a mixed number or whole number.**

  Divide the denominator into the numerator.

  $\frac{31}{4} = 7\frac{3}{4}$

$$\begin{array}{r} 7\frac{3}{4} \\ 4\ \overline{)31} \\ -28 \\ \hline 3 \end{array}$$

16 **Add or Subtract fractions**-to add or subtract fractions the denominators have to be the same because the pieces have to be the same size to add to what you start with or to subtract from it.

- If given $\frac{2}{7}+\frac{3}{7}=\frac{5}{7}$ $\quad \frac{5}{6}-\frac{2}{6}=\frac{3}{6}=\frac{1}{2}$

$$\frac{1}{3}+\frac{1}{3}=\frac{2}{3} \qquad \frac{7}{9}-\frac{4}{9}=\frac{3}{9}=\frac{1}{3}$$

BUT.......most the time the denominators will not be the same. Here is a one handed way to add or subtract fractions that do not have the same denominators!

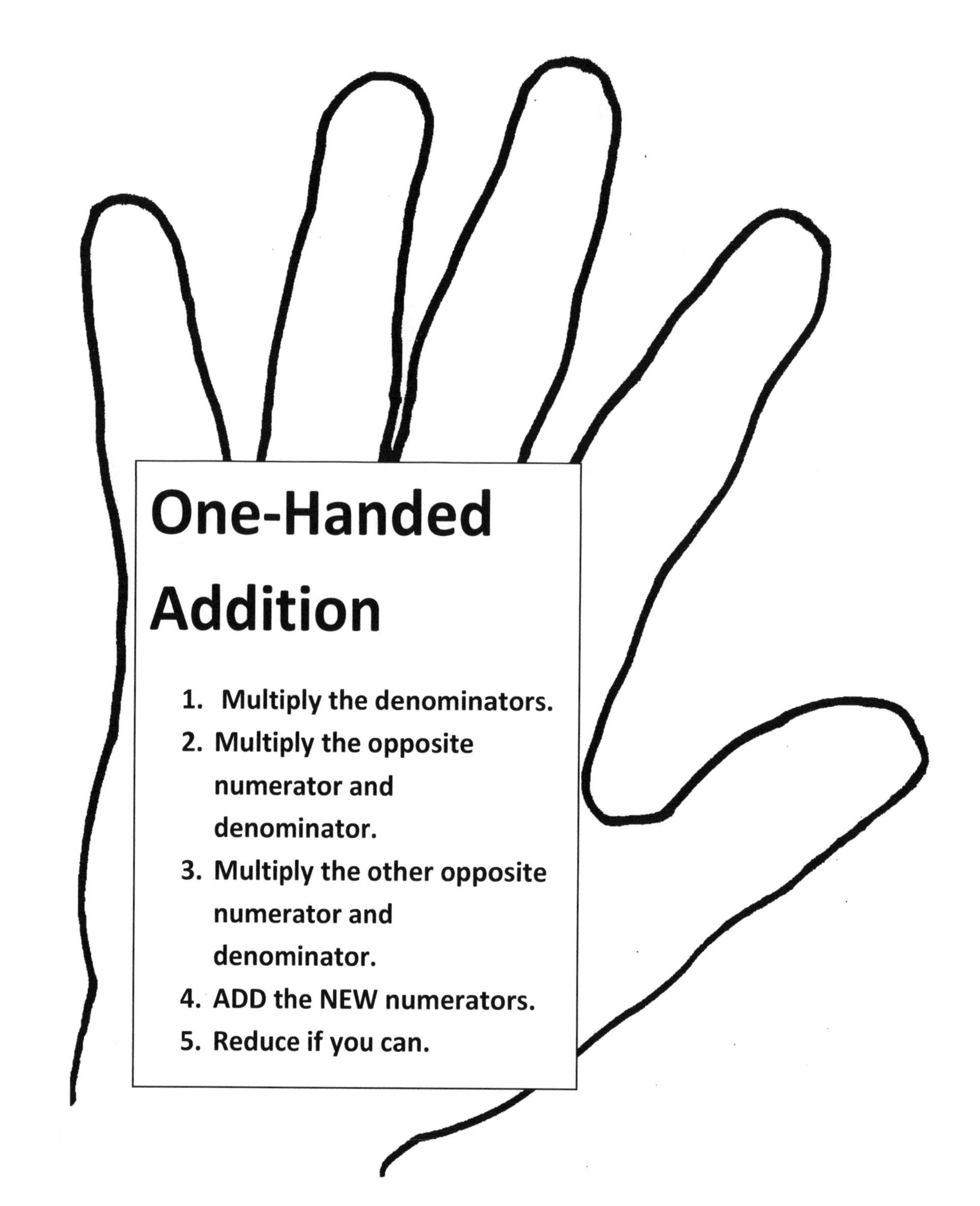
One-Handed Addition
1. Multiply the denominators.
2. Multiply the opposite numerator and denominator.
3. Multiply the other opposite numerator and denominator.
4. ADD the NEW numerators.
5. Reduce if you can.

Steps:

1. 2 * 3 = 6

1. Multiply the denominators.

$$\frac{1}{2} + \frac{1}{3} = \frac{}{6}$$

2. 1 * 3 = 3

2. Multiply one opposite numerator and denominator.

3

$$\frac{1}{2} + \frac{1}{3} = \frac{}{6}$$

3. 1 * 2 = 2

3. Multiply the other opposite numerator and denominator.

2

$$\frac{1}{2} + \frac{1}{3} = \frac{}{6}$$

4. 3 + 2 = 5

4. Add the new numerators.

3 + 2

$$\frac{1}{2} + \frac{1}{3} = \frac{5}{6}$$

5. There is no more reducing that can be done.

5. Reduce if you can.

**Remember there will be many times when you will leave an improper fraction, but it must be reduced

15 + 8

1. $\frac{3}{4} + \frac{2}{5} = \frac{23}{20} \quad or \quad 1\frac{3}{20}$

15 + 6

2. $\frac{5}{6} + \frac{1}{3} = \frac{21}{18} = \frac{7}{6} \quad or \quad 1\frac{1}{6}$

24+45

3. $\frac{4}{9} + \frac{5}{6} = \frac{69}{54} \div \frac{3}{3} = \frac{23}{18} \quad or \quad 1\frac{5}{18}$

Resist the trap of over reducing!

30+14

4. $\frac{6}{7} + \frac{2}{5} = \frac{44}{35} \quad or \quad 1\frac{9}{35}$

These fractions have factors, but no common ones, so they cannot be reduced.

Problems for you to work.

1. $\frac{3}{4} + \frac{1}{3} =$

2. $\frac{5}{8} + \frac{1}{4} =$

3. $\frac{2}{5} + \frac{2}{3} =$

4. $\frac{2}{9} + \frac{5}{6} =$

5. $\frac{2}{3} + \frac{5}{6} =$

6. $\frac{3}{4} + \frac{1}{2} =$

Check Your Answers

1. $\overset{9 + 4}{\frac{3}{4}+\frac{1}{3}}=\frac{13}{12} \quad or \quad 1\frac{1}{12}$

2. $\overset{20+ 8}{\frac{5}{8}+\frac{1}{4}}=\frac{28}{32}=\frac{7}{8}$

3. $\overset{6 +10}{\frac{2}{5}+\frac{2}{3}}=\frac{16}{15} \quad or \quad 1\frac{1}{15}$

4. $\overset{12+45}{\frac{2}{9}+\frac{5}{6}}=\frac{57}{54}\div\frac{3}{3}=\frac{19}{18} \quad or \quad 1\frac{1}{18}$

5. $\overset{12+15}{\frac{2}{3}+\frac{5}{6}}=\frac{27}{18}\div\frac{9}{9}=\frac{3}{2} \quad or \quad 1\frac{1}{2}$

6. $\overset{6 + 4}{\frac{3}{4}+\frac{1}{2}}=\frac{10}{8}\div\frac{2}{2}=\frac{5}{4} \quad or \quad 1\frac{1}{4}$

Skill Stop

If you are having some problems with reducing fractions, try using the divisibility rules.

1. If both fractions are even, they will divide by **2.**
$\frac{16}{24}$ both fractions are even, so they will divide by 2.
$\frac{16}{24} \div \frac{2}{2} = \frac{8}{12}$ Here you still have two even fractions, so divide by 2 again, or if you see it divide them both by 4. ***Sometimes once you have divided the two fractions one time, you will see another common factor.***

2. Numbers divide by 3 if the sum of the digits divide by **3.** $\frac{54}{72}$ $5 + 4 = 9,\ 9 \div 3, so\ 54 \div 3\ 7 + 2 = 9, 9 \div 3, so\ 72 \div 3$ Since both numbers of the fractions divide by 3, that is a good place to start, or you can divide by 9 if you see that and not have to reduce the fraction twice by 3. $\frac{54}{72} \div \frac{3}{3} = \frac{18}{21} \div \frac{3}{3} = \frac{6}{7}$

3. By far the easiest divisibility rule to use is **5**. If both fractions end in 5, the fraction will reduce by 5. If both end in 0, the fraction will reduce by 5. If one fraction ends in a 5 and the other in a 0, the fraction will reduce by 5. $\frac{40}{55} \div \frac{5}{5} = \frac{8}{11}$

4. Another way to help figure out what to reduce by is to **skip count.** If both numbers of the fractions show up in the same line, they will reduce by the line leader.

2, 4, 6, 8, 10, 12, 14, 16, 18, 20, 22, 24
3, 6, 9, 12, 15, 18, 21, 24, 27, 30, 33, 36
4, 8, 12, 16, 20, 24, 28, 32, 36, 40, 44, 48
5, 10, 15, 20, 25, 30, 35, 40, 45, 50, 55, 60
6, 12, 18, 24, 30, 36, 42, 48, 54, 60, 66, 72
7, 14, 21, 28, 35, 42, 49, 56, 63, 70, 77, 84
8, 16, 24, 32, 40, 48, 56, 64, 72, 80, 88, 96
9, 18, 27, 36, 45, 54, 63, 72, 81, 90, 99, 108

In a nutshell, if you are having a problem reducing try:

1. Are they both even? Divide each by 2
2. Won't divide by 2? Add the digits, if their sum divides by 3, the fraction reduces by 3.
3. Do the numbers of your fraction end in a 5 or a 0? Reduce the fraction by 5.
4. Try looking for both numbers of the fraction in the same skip count line. If they are in the same line, reduce by the line leader.

Remember it does not matter how many times you have to reduce a fraction, just make sure you leave it in simplest terms or as small as you can without over reducing.

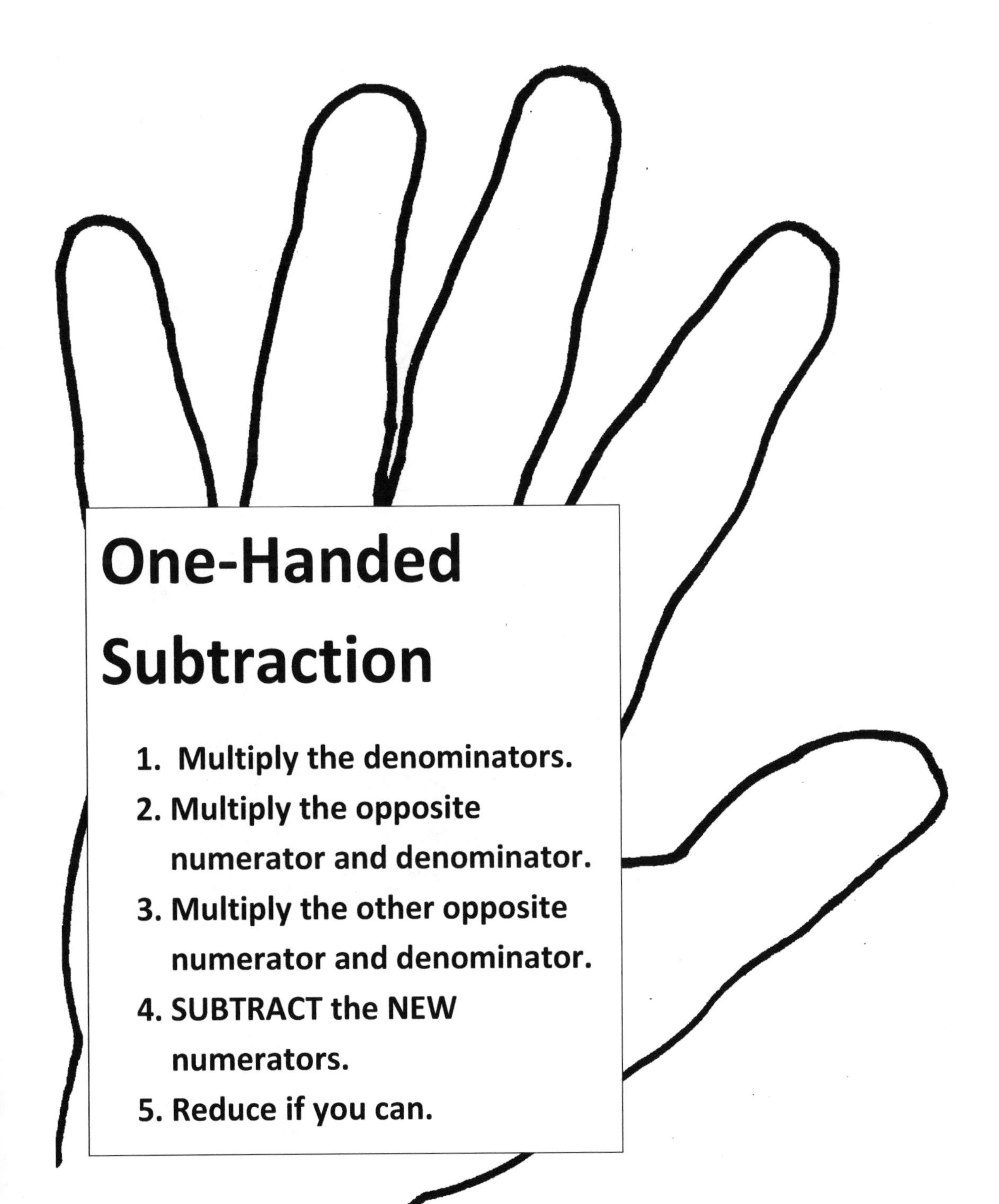

One-Handed Subtraction

1. Multiply the denominators.
2. Multiply the opposite numerator and denominator.
3. Multiply the other opposite numerator and denominator.
4. SUBTRACT the NEW numerators.
5. Reduce if you can.

**Subtraction of fractions works the exact same way as adding fractions except you subtract the new numerators instead of adding them.

**Subtraction is easier than addition because there is much less reducing, but be careful because there is some.

Examples

9 - 5

1. $\frac{3}{5} - \frac{1}{3} = \frac{4}{15}$

15 - 6

2. $\frac{5}{6} - \frac{1}{3} = \frac{9}{18} = \frac{1}{2}$

Steps

You should be ready for the short version.

1. Multiply.
2. Multiply.
3. Multiply.
4. Add or subtract the numerator.
5. Reduce if you can.

**You will find that you make fewer mistakes if you put the subtraction or addition sign up between the two new numerators.

Subtracting Fractions

Examples

9 - 5

1. $\frac{3}{5} - \frac{1}{3} = \frac{4}{15}$

15 -6

2. $\frac{5}{6} - \frac{1}{3} = \frac{9}{18} = \frac{1}{2}$

6 - 10

3. $\frac{1}{2} - \frac{5}{6} = -\frac{4}{12} = -\frac{1}{3}$

Resist the trap of over reducing!

20-14

4. $\frac{4}{7} - \frac{2}{5} = \frac{6}{35}$

Theses numbers have factors, but none in common.

Problems for you to work.

1. $\frac{5}{6} - \frac{1}{2} =$

2. $\frac{3}{7} - \frac{1}{3} =$

3. $\frac{7}{9} - \frac{2}{3} =$

4. $\frac{4}{5} - \frac{3}{4} =$

5. $\frac{1}{2} - \frac{3}{5} =$

6. $\frac{2}{3} - \frac{1}{6} =$

Check your answers.

10 - 6

1. $\frac{5}{6} - \frac{1}{2} = \frac{4}{12} = \frac{1}{3}$

9 - 7

2. $\frac{3}{7} - \frac{1}{3} = \frac{2}{21}$

21 - 18

3. $\frac{7}{9} - \frac{2}{3} = \frac{3}{27} = \frac{1}{9}$

16 - 15

4. $\frac{4}{5} - \frac{3}{4} = \frac{1}{20}$

5 - 6

5. $\frac{1}{2} - \frac{3}{5} = -\frac{1}{10}$

12 - 3

6. $\frac{2}{3} - \frac{1}{6} = \frac{9}{18} - \frac{1}{2}$

*****Here are some helpful hints to remember at this point.**

1. Say the steps to yourself as you do the problems and it will help them stick in your mind sooner.

2. ALWAYS divide the numerator and the denominator by the same number when reducing your answer.

3. Do not worry about how strange your answer looks as long as you have followed the steps and you have not made any calculation mistakes.

4. Make sure you take a second look at your answer to be sure you cannot reduce it, but do not over reduce. Every answer will not reduce.

5. Make sure you have taken care of any improper fractions. Remember in higher math, you will mostly leave the fraction as improper, but you must reduce it if you can.

6. As noted earlier, there is a lot less reducing in subtraction of fractions than in addition, but look to make sure.

7. Bring up the sign between your new numerators when adding or subtracting fractions.

8. Don't be afraid of fractions-----follow the steps and you **WILL** get them correct!!!!

Addition and Subtraction of Fractions Practice

1. $\frac{6}{7} - \frac{3}{4} =$

2. $\frac{5}{9} + \frac{1}{2} =$

3. $\frac{5}{6} + \frac{2}{3} =$

4. $\frac{2}{3} - \frac{4}{5} =$

5. $\frac{4}{7} - \frac{1}{2} =$

6. $\frac{4}{5} + \frac{4}{9} =$

7. $\frac{11}{12} - \frac{5}{6} =$

8. $\frac{9}{10} + \frac{3}{5} =$

9. $\frac{1}{2} - \frac{6}{7} =$

10. $\frac{3}{7} + \frac{9}{10} =$

Check Your Answers

1. $\overset{24-21}{}\quad \frac{6}{7}-\frac{3}{4}=\frac{3}{28}$

2. $\overset{10+8}{}\quad \frac{5}{9}+\frac{1}{2}=\frac{19}{18} \quad or \quad 1\frac{1}{18}$

3. $\overset{15+12}{}\quad \frac{5}{6}+\frac{2}{3}=\frac{27}{18}=\frac{3}{2} \quad or \quad 1\frac{1}{2}$

4. $\overset{10-12}{}\quad \frac{2}{3}-\frac{4}{5}=-\frac{2}{15}$

5. $\overset{8-7}{}\quad \frac{4}{7}-\frac{1}{2}=\frac{1}{14}$

6. $\overset{36+20}{}\quad \frac{4}{5}+\frac{4}{9}=\frac{56}{45} \quad or \quad 1\frac{11}{45}$

7. $\overset{66-60}{}\quad \frac{11}{12}-\frac{5}{6}=\frac{6}{72}=\frac{1}{12}$

45+30

8. $\frac{9}{10}+\frac{3}{5}=\frac{75}{50}=\frac{15}{10}=\frac{3}{2}$ $\quad or \quad$ $1\frac{1}{2}$

7-12

9. $\frac{1}{2}-\frac{6}{7}= -\frac{5}{14}$

30+63

10. $\frac{3}{7}+\frac{9}{10}=\frac{93}{70}$ $\quad or \quad$ $1\frac{23}{70}$

SKILL STOP

- **ALWAYS** change a mixed number or whole number to an improper fraction **before** adding, subtracting, multiplying, or dividing.

- If you have forgotten how to change to an improper fraction, look back on page 7.

- After changing to an improper fraction follow the 5 step, One-Handed method for that operation.

One-Handed Addition of Whole and Mixed Numbers

***Remember these problems have a step before you start the One-Handed Fractions steps.

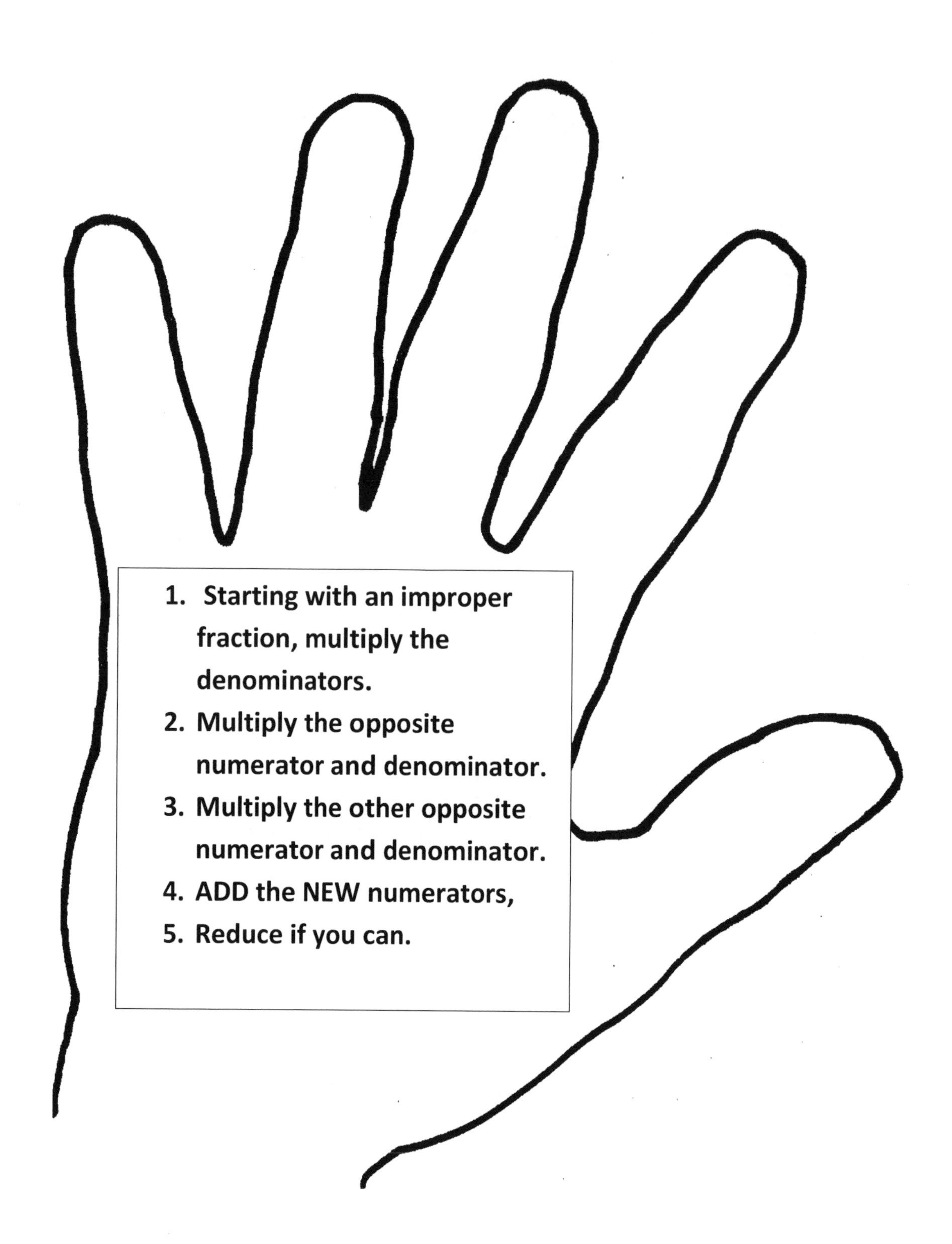
1. Starting with an improper fraction, multiply the denominators.
2. Multiply the opposite numerator and denominator.
3. Multiply the other opposite numerator and denominator.
4. ADD the NEW numerators,
5. Reduce if you can.

These fractions need an extra step because the first thing you will do regardless of the sign, is change the whole or mixed number to an improper fraction.

Example

$$\mathbf{2\frac{1}{3} + 1\frac{5}{6} = \frac{7}{3} + \frac{11}{6} =}$$

$$\mathbf{2\frac{1}{3} = 2 * 3 = 6 + 1 = \frac{7}{3}}$$

$$\mathbf{1\frac{5}{6} = 1 * 6 = 6 + 5 = \frac{11}{6}}$$

Once you have changed to an improper fraction, you follow the steps that have already been laid out for you to add and subtract fractions.

$$42 + 33$$

1. Multiply $2\frac{1}{3} + 1\frac{5}{6} = \frac{7}{3} + \frac{11}{6} = \frac{75}{18} = \frac{25}{6} \quad or \; 4\frac{1}{6}$
2. Multiply
3. Multiply
4. Add or Subtract
5. Reduce if you can

Examples

$$154+65$$

1. $4\frac{2}{5}+1\frac{6}{7}=\frac{22}{5}+\frac{13}{7}=\frac{219}{35} \quad or \quad 6\frac{9}{35}$

$$66- 33$$

2. $3\frac{2}{3}-1\frac{5}{6}=\frac{11}{3}-\frac{11}{6}=\frac{33}{18}=\frac{11}{6} \quad or \quad 1\frac{5}{6}$

$$100 + 63$$

3. $2\frac{7}{9}+1\frac{3}{4}=\frac{25}{9}+\frac{7}{4}=\frac{163}{36} \quad or \quad 4\frac{19}{36}$

$$40-28$$

4. $1\frac{1}{4}-\frac{7}{8}=\frac{5}{4}-\frac{7}{8}=\frac{12}{32}=\frac{3}{8}$

- Do not worry if you have to reduce more than once.
- Also, remember there are times to stop with a reduced, improper fraction.

** It is easy to forget you have not done the multiplication of the two opposite numerator/denominator pairs when you have mixed numbers because the improper fractions are pretty big sometimes. You may have the temptation to just add the two numerators of the improper fractions. Just say the steps to yourself as you do them until the steps become second nature.

1. Multiply.
2. Multiply.
3. Multiply.
4. Add the **NEW** numerators.
5. Reduce if you can.

***Also, remember that in higher math, you will usually leave the improper fraction as your answer.

Problems for you to work.

1. $1\frac{3}{5} + 2\frac{1}{3} =$

2. $3\frac{1}{3} + 1\frac{4}{9} =$

3. $\frac{8}{9} + 2\frac{2}{3} =$

4. $4\frac{1}{2} + 1\frac{5}{6} =$

5. $1\frac{3}{10} + 2\frac{3}{5} =$

6. $2\frac{3}{4} + 1\frac{7}{8} =$

Check your to work.

1. $$\overset{24+35}{}$$ $1\frac{3}{5}+2\frac{1}{3}=\frac{8}{5}+\frac{7}{3}=\frac{59}{15}$ *or* $3\frac{14}{15}$

2. $$\overset{90\ +\ 39}{}$$ $3\frac{1}{3}+1\frac{4}{9}=\frac{10}{3}+\frac{13}{9}=\frac{129}{27}=\frac{43}{9}$ *or* $4\frac{7}{9}$

3. $$\overset{24+72}{}$$ $\frac{8}{9}+2\frac{2}{3}=\frac{8}{9}+\frac{8}{3}=\frac{96}{27}=\frac{32}{9}$ *or* $3\frac{5}{9}$

4. $$\overset{54+22}{}$$ $4\frac{1}{2}+1\frac{5}{6}=\frac{9}{2}+\frac{11}{6}=\frac{76}{12}=\frac{19}{3}$ *or* $6\frac{1}{3}$

5. $$\overset{65+130}{}$$ $1\frac{3}{10}+2\frac{3}{5}=\frac{13}{10}+\frac{13}{5}=\frac{195}{50}=\frac{39}{10}$ *or* $3\frac{9}{10}$

6. $$\overset{88+60}{}$$ $2\frac{3}{4}+1\frac{7}{8}=\frac{11}{4}+\frac{15}{8}=\frac{148}{32}=\frac{37}{8}$ *or* $4\frac{5}{8}$

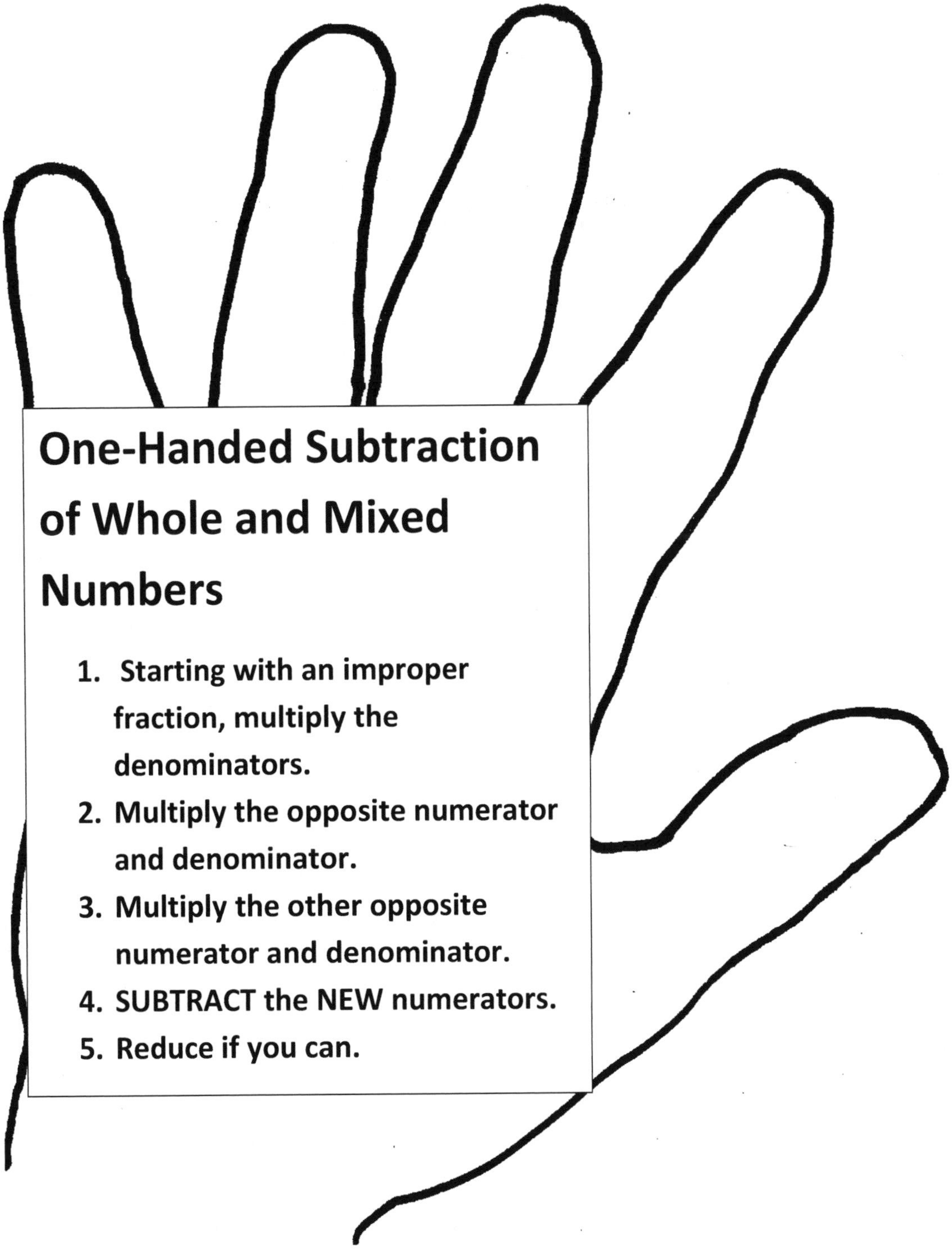

One-Handed Subtraction of Whole and Mixed Numbers

1. Starting with an improper fraction, multiply the denominators.
2. Multiply the opposite numerator and denominator.
3. Multiply the other opposite numerator and denominator.
4. SUBTRACT the NEW numerators.
5. Reduce if you can.

Subtraction of Mixed Numbers and Whole Numbers is the same as addition. Once you change the Mixed Numbers or Whole Numbers to an improper fraction, you follow the 5 steps for subtracting fractions.

1. Multiply.
2. Multiply.
3. Multiply.
4. Subtract the **NEW** numerators.
5. Reduce if you can.

Example

70 - 33

$$3\frac{1}{3} - 1\frac{4}{7} = \frac{10}{3} - \frac{11}{7} = \frac{37}{21} \quad or \quad 1\frac{16}{21}$$

***Be careful with fractions like $\frac{16}{21}$ as they look like they will reduce, but these do not have any common factors.

Problems for You to work

1. $5\frac{1}{3} - 3\frac{5}{6} =$

2. $4 - 2\frac{2}{3} =$

3. $2\frac{5}{7} - 1\frac{3}{4} =$

4. $1\frac{1}{2} - \frac{8}{9} =$

5. $7\frac{1}{2} - 4\frac{5}{6} =$

6. $3\frac{2}{3} - 1\frac{1}{2} =$

Check your work

96-69

1. $5\frac{1}{3} - 3\frac{5}{6} = \frac{16}{3} - \frac{23}{6} = \frac{27}{18} = \frac{3}{2} \quad or \quad 1\frac{1}{2}$

12-8

2. $4 - 2\frac{2}{3} = \frac{4}{1} - \frac{8}{3} = \frac{4}{3} \quad or \quad 1\frac{1}{3}$

76-49

3. $2\frac{5}{7} - 1\frac{3}{4} = \frac{19}{7} - \frac{7}{4} = \frac{27}{28}$

27- 16

4. $1\frac{1}{2} - \frac{8}{9} = \frac{3}{2} - \frac{8}{9} = \frac{11}{18}$

90-58

5. $7\frac{1}{2} - 4\frac{5}{6} = \frac{15}{2} - \frac{29}{6} = \frac{32}{12} = \frac{8}{3} \quad or \quad 2\frac{2}{3}$

22-9

6. $3\frac{2}{3} - 1\frac{1}{2} = \frac{11}{3} - \frac{3}{2} = \frac{13}{6} \quad or \quad 2\frac{1}{6}$

Mixed addition and subtract problems for you to work.

1. $4\frac{1}{3} - 3\frac{1}{2} =$

2. $1\frac{2}{3} + 1\frac{4}{5} =$

3. $2\frac{1}{3} + 1\frac{7}{8} =$

4. $3\frac{5}{6} - 1\frac{2}{3} =$

5. $3\frac{1}{4} - \frac{7}{9} =$

6. $2\frac{3}{4} + 2\frac{1}{2} =$

Check your to work.

26-21

1. $4\frac{1}{3} - 3\frac{1}{2} = \frac{13}{3} - \frac{7}{2} = \frac{5}{6}$

25+27

2. $1\frac{2}{3} + 1\frac{4}{5} = \frac{5}{3} + \frac{9}{5} = \frac{52}{15} \quad or \quad 3\frac{7}{15}$

56+45

3. $2\frac{1}{3} + 1\frac{7}{8} = \frac{7}{3} + \frac{15}{8} = \frac{101}{24} \quad or \quad 4\frac{5}{24}$

69-30

4. $3\frac{5}{6} - 1\frac{2}{3} = \frac{23}{6} - \frac{5}{3} = \frac{39}{18} = \frac{13}{6} \quad or \quad 2\frac{1}{6}$

117-28

5. $3\frac{1}{4} - \frac{7}{9} = \frac{13}{4} - \frac{7}{9} = \frac{89}{36} \quad or \quad 2\frac{17}{36}$

22+20

6. $2\frac{3}{4} + 2\frac{1}{2} = \frac{11}{4} + \frac{5}{2} = \frac{42}{8} = \frac{21}{4} \quad or \quad 5\frac{1}{4}$

Addition and Subtraction of Fractions

- Remember that fractions cannot be added or subtracted unless they are in the same size piece. This means that their denominators are the same.
- Unfortunately, the denominators are rarely the same when you need to add or subtract the fractions.
- The One-Handed method will work quickly on any add or subtraction of fractions problem. There are other ways, but trust me, this is by far the easiest way!
- Say the step to yourself for several problems and they will become second nature to you.
    1. Multiply.
    2. Multiply.
    3. Multiply.
    4. Add or subtract the numerators.
    5. Reduce if you can.

Hold

Everything!!

The RULES are a CHANGING!

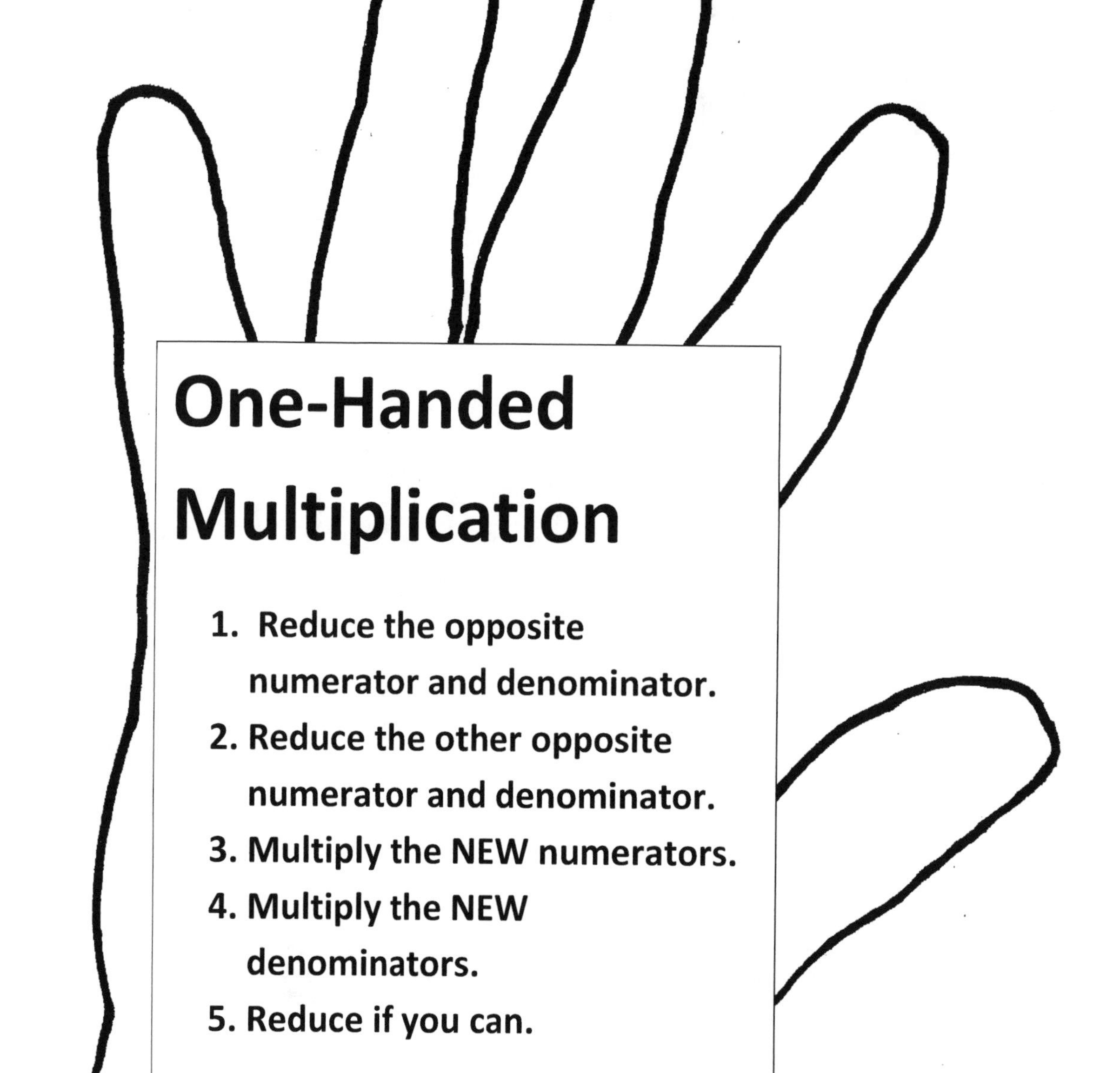
One-Handed Multiplication
1. Reduce the opposite numerator and denominator.
2. Reduce the other opposite numerator and denominator.
3. Multiply the NEW numerators.
4. Multiply the NEW denominators.
5. Reduce if you can.

Multiplying Fractions

This is my favorite type of fraction problem to do, because they are so easy. The reducing is done **before** the multiplying is done! The X sign is the **only** sign that reducing can be done across. I know in higher math we use the * sign for multiplication, but you can think back to grade school math in your head and the X sign points to the two numbers that can be reduced first. Of course sometimes you cannot reduce at all or may only be able to reduce one pair. When you can reduce both pairs, the problem is done in a snap. Just remember this only happens when we multiply fractions.

Example: $\frac{3}{5}\,X\,\frac{7}{9} =$

1. 3 and 9 are one opposite numerator/denominator pair. They both divide by 3. $3 \div 3 = 1$ $\quad 9 \div 3 = 3$, so now you have $\frac{1}{5}\,x\,\frac{7}{3} =$.
2. 5 and 7 are the other opposite numerator/denominator-pair. These have no common factors, so you still have $\frac{1}{5}\,x\,\frac{7}{3} =$.

3. Multiply the numerators $\frac{1x7}{5x3} = \frac{7}{5x3}$
4. Multiply the denominators $\frac{7}{5x3} = \frac{7}{15}$
5. Reduce if you can. There is no more reducing that can be done, so you are finished.

I know it can be confusing how you can multiply two fractions and end up with a fraction that is smaller than you started with. This does not happen when you multiply whole numbers.

With multiplication of fractions, you are taking a part of a whole like $\frac{1}{2}$ and finding $\frac{1}{4}$ of it, which would only be $\frac{1}{8}$.

The short way to these problems is like this:

$$\frac{\overset{1}{\cancel{4}}}{\underset{3}{\cancel{9}}} \, x \, \frac{\overset{1}{\cancel{3}}}{\underset{5}{\cancel{20}}} = \frac{1}{15}$$

STEPS

1. Reduce opposites.
2. Reduce opposites.
3. Multiply numerators.
4. Multiply denominators.
5. Reduce if you can.

1. $\frac{5}{8}\ x\ \frac{4}{5} = \quad 5 \div 5 = 1;\ \ 4 \div 4 = 1\ and\ \ 8\ \div 4 = 2\ \ so$

$$\frac{\overset{1}{\cancel{5}}}{\underset{2}{\cancel{8}}}\ x\ \frac{\overset{1}{\cancel{4}}}{\underset{1}{\cancel{5}}} = \frac{1x1}{2x1} = \frac{1}{2}$$

2. $\frac{2}{3}\ x\ \frac{6}{7} = \ \ 2\ and\ 7\ do\ not\ reduce.\ 3 \div 3 = 1\ and\ 6 \div 3 = 2\ \ so$

$$\frac{2}{\underset{1}{\cancel{3}}}\ x\ \frac{\overset{2}{\cancel{6}}}{7} = \frac{2x2}{1x7} = \frac{4}{7}$$

3. $\frac{4}{9}\ x\ \frac{3}{4} = \quad 4 \div 4 = 1\ and\ \ 3\ \div 3 = 1\ and\ \ 9 \div 3 = 3\ \ \ so$

$$\frac{\overset{1}{\cancel{4}}}{\underset{3}{\cancel{9}}}\ x\ \frac{\overset{1}{\cancel{3}}}{\underset{1}{\cancel{4}}} = \frac{1x1}{3x1} = \frac{1}{3}$$

4. $\frac{9}{10}\ x\ \frac{10}{27} \ \ = \ \ 9 \div 9 = 1\ and\ 27\ \ \div 9 = 3;\ \ 10\ \ \div 10 = 1\ \ so$

$$\frac{\overset{1}{\cancel{9}}}{\underset{1}{\cancel{10}}}\ x\ \frac{\overset{1}{\cancel{10}}}{\underset{3}{\cancel{27}}} = \frac{1x1}{1x3} = \frac{1}{3}$$

***Usually with multiplication we do not use this many steps. When you are ready, you will leave out the extra step. $\frac{1x1}{1x3}$

You try some.

1. $\frac{6}{7}\ x\frac{7}{9} =$
2. $\frac{13}{14}\ x\frac{7}{9} =$
3. $\frac{1}{2}\ x\frac{4}{7} =$
4. $\frac{3}{5}\ x\frac{10}{27} =$
5. $\frac{11}{12}\ x\frac{4}{5} =$
6. $\frac{9}{10}\ x\frac{5}{9} =$
7. $\frac{1}{2}\ x\frac{15}{16} =$
8. $\frac{3}{7}\ x\frac{14}{15} =$
9. $\frac{8}{9}\ x\frac{3}{4} =$
10. $\frac{7}{10}\ x\frac{5}{49} =$

Check your answers.

1. $\frac{{}^{2}\cancel{6}}{{}_{1}\cancel{7}} \; x \; \frac{{}^{1}\cancel{7}}{{}_{3}\cancel{9}} = \frac{2x1}{1x3} = \frac{2}{3}$

2. $\frac{13}{{}_{2}\cancel{14}} \; x \; \frac{{}^{1}\cancel{7}}{9} = \frac{13x1}{2x9} = \frac{13}{18}$

3. $\frac{1}{{}_{1}\cancel{2}} \; x \; \frac{{}^{2}\cancel{4}}{7} = \frac{1x2}{1x7} = \frac{2}{7}$

4. $\frac{{}^{1}\cancel{3}}{{}_{1}\cancel{5}} \; x \; \frac{{}^{2}\cancel{10}}{{}_{9}\cancel{27}} = \frac{1x2}{1x9} = \frac{2}{9}$

5. $\frac{11}{{}_{3}\cancel{12}} \; x \; \frac{{}^{1}\cancel{4}}{5} = \frac{11x1}{3x5} = \frac{11}{15}$

6. $\frac{{}^{1}\cancel{9}}{{}_{2}\cancel{10}} \; x \; \frac{{}^{1}\cancel{5}}{{}_{1}\cancel{9}} = \frac{1x1}{2x1} = \frac{1}{2}$

7. $\frac{1}{2} \; x \; \frac{15}{16} = \frac{1x15}{2x16} = \frac{15}{32}$

8. $\frac{\overset{1}{\cancel{3}}}{\underset{1}{\cancel{7}}} \ x \ \frac{\overset{2}{\cancel{14}}}{\underset{5}{\cancel{15}}} = \frac{1x2}{1x5} = \frac{2}{5}$

9. $\frac{\overset{2}{\cancel{8}}}{\underset{3}{\cancel{9}}} \ x \ \frac{\overset{1}{\cancel{3}}}{\underset{1}{\cancel{4}}} = \frac{2x1}{3x1} = \frac{2}{3}$

10. $\frac{\overset{1}{\cancel{7}}}{\underset{2}{\cancel{10}}} \ x \ \frac{\overset{1}{\cancel{5}}}{\underset{7}{\cancel{49}}} = \frac{1x1}{2x7} = \frac{1}{14}$

If you do not need the extra step, leave it off. You do not need it unless it helps you to see what you are multiplying after the reducing is finished.

Now try some mixed +, -, and x practice.

1. $\frac{7}{8} - \frac{4}{5} =$

2. $\frac{5}{9}\ x\,\frac{3}{4} =$

3. $\frac{9}{10} + \frac{5}{6} =$

4. $\frac{9}{10}\ x\,\frac{5}{6} =$

5. $\frac{3}{4} + \frac{8}{9} =$

6. $\frac{4}{5} - \frac{1}{3} =$

7. $\frac{8}{9}\ x\,\frac{15}{16} =$

8. $\frac{10}{11}\ x\,\frac{11}{12} =$

9. $\frac{3}{5} + \frac{2}{3} =$

10. $\frac{7}{9} - \frac{2}{3} =$

Check your answers.

35 - 32

1. $\frac{7}{8} - \frac{4}{5} = \frac{3}{40}$

2. $\frac{5}{\cancel{9}_3} \; x \; \frac{\cancel{3}^1}{4} = \frac{5x1}{3x4} = \frac{5}{12}$

54 + 50

3. $\frac{9}{10} + \frac{5}{6} = \frac{104}{60} = \frac{26}{15} \quad or \quad 1\frac{11}{15}$

4. $\frac{\cancel{9}^3}{\cancel{10}_2} \; x \; \frac{\cancel{5}^1}{\cancel{6}_2} = \frac{3x1}{2x2} = \frac{3}{4}$

27+32

5. $\frac{3}{4} + \frac{8}{9} = \frac{59}{36} \quad or \quad 1\frac{23}{36}$

12 - 5

6. $\frac{4}{5} - \frac{1}{3} = \frac{7}{15}$

7. $\frac{\cancel{8}^1}{\cancel{9}_3} \; x \; \frac{\cancel{15}^5}{\cancel{16}_2} = \frac{1x5}{3x2} = \frac{5}{6}$

8. $\frac{\cancel{10}^5}{\cancel{11}_1} \; x \; \frac{\cancel{11}^1}{\cancel{12}_6} = \frac{5x1}{1x6} = \frac{5}{6}$

9 + 10

9. $\frac{3}{5}+\frac{2}{3}=\frac{19}{15} \quad or \quad 1\frac{4}{15}$

21-18

10. $\frac{7}{9}-\frac{2}{3}=\frac{3}{27}=\frac{1}{9}$

***Remember addition and subtraction of fractions use the same 5 steps:

1. Multiply.
2. Multiply.
3. Multiply.
4. Add or Subtract the numerators.
5. Reduce if you can.

***You cannot reduce across an addition or subtraction sign!

***Remember to bring the addition or subtraction sign up between the new numerators so you remember whether to add or subtract them.

***Multiplication follows its own rules.

1. Reduce opposites.
2. Reduce other opposites.
3. Multiply the numerators.
4. Multiply the denominators.
5. Reduce if you can. ***This is usually only needed if you missed a reduction before you multiplied.

***Don 't over reduce.

SKILL STOP

- **ALWAYS change a mixed number or whole number to an improper fraction before adding, subtracting, multiplying, or dividing.**

- If you have forgotten how to change to improper fractions, look back on page 7.

- **After changing to an improper fraction follow the 5 step, One-Handed method for that operation.**

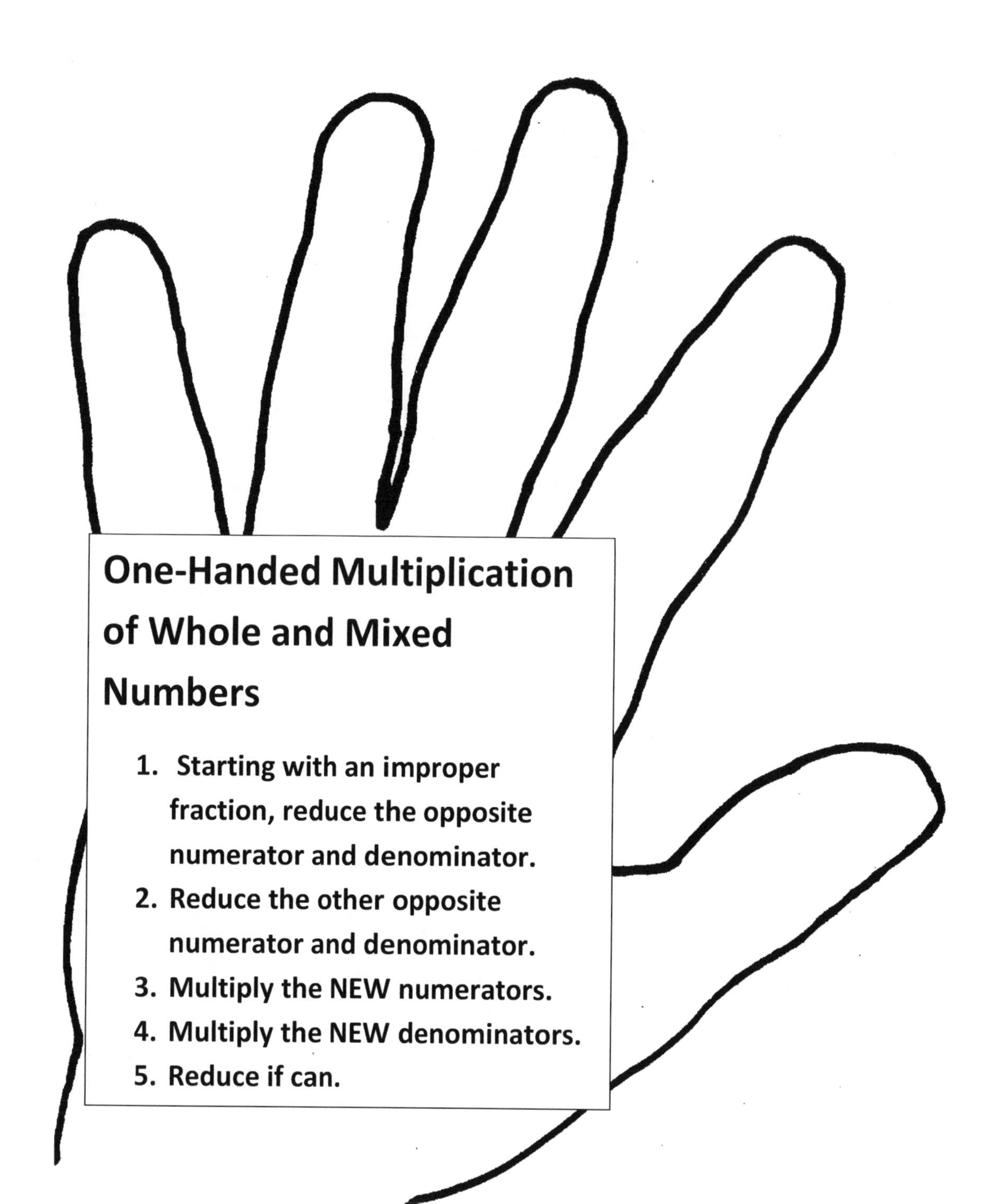

One-Handed Multiplication of Whole and Mixed Numbers

1. Starting with an improper fraction, reduce the opposite numerator and denominator.
2. Reduce the other opposite numerator and denominator.
3. Multiply the NEW numerators.
4. Multiply the NEW denominators.
5. Reduce if can.

Multiplication of Mixed Numbers

***Like addition and subtraction of mixed numbers and whole numbers, the first step is to turn the mixed number or whole number to an improper fraction.

Then just like the other operations, you will follow the same 5 step, One-Handed method that you used for regular fractions with that operation.

For Multiplication:

1. Reduce.
2. Reduce.
3. Multiply.
4. Multiply.
5. Reduce if you can.

***There will be more reducing with mixed numbers and whole numbers because sometimes you will change the improper fraction answer back into a whole number or mixed number.

***But, also remember with higher math, you will leave the improper fraction as your answer.

Examples

1. $1\frac{4}{5}\ x\ 1\frac{2}{3} = \frac{\overset{3}{\cancel{9}}}{\underset{1}{\cancel{5}}}\ x\ \frac{\overset{1}{\cancel{5}}}{\underset{1}{\cancel{3}}} = \frac{3x1}{1x1} = \frac{3}{1} \quad or \quad 3$

2. $1\frac{1}{2}\ x\ 2\frac{1}{6} = \frac{\overset{1}{\cancel{3}}}{2}\ x\ \frac{13}{\underset{2}{\cancel{6}}} = \frac{1x13}{2x2} = \frac{13}{4} \quad or \quad 3\frac{1}{4}$

***I will leave out the extra step of rewriting the problem once you have done the reducing. If you need it, keep it.

3. $2\frac{2}{5}\ x\ 1\frac{7}{8} = \frac{\overset{3}{\cancel{12}}}{\underset{1}{\cancel{5}}}\ x\ \frac{\overset{3}{\cancel{15}}}{\underset{2}{\cancel{8}}} = \frac{9}{2} \quad or \quad 4\frac{1}{2}$

4. $3\frac{1}{3}\ x\ 1\frac{4}{5} = \frac{\overset{2}{\cancel{10}}}{\underset{1}{\cancel{3}}}\ x\ \frac{\overset{3}{\cancel{9}}}{\underset{1}{\cancel{5}}} = \frac{6}{1} \quad or \quad 6$

Practice Problems

1. $2\frac{2}{3}\ x\ 1\frac{5}{16} =$

2. $3\frac{3}{5}\ x\ 2\frac{2}{9} =$

3. $6\frac{1}{2}\ x\ 2\frac{2}{5} =$

4. $6\ x\ 4\frac{5}{6} =$

5. $4\frac{2}{3}\ x\ 1\frac{5}{7} =$

6. $1\frac{7}{8}\ x\ 1\frac{3}{7} =$

Check Your Answers.

1. $2\frac{2}{3} \ x \ 1\frac{5}{16} = \frac{8}{3} \ x \ \frac{21}{16} = \frac{7}{2} \ or \ 3\ ½$

2. $3\frac{3}{5} \ x \ 2\frac{2}{9} = \frac{18}{5} \ x \ \frac{20}{9} = \frac{8}{1} \ or \ 8$

3. $6\frac{1}{2} \ x \ 2\frac{2}{5} = \frac{13}{2} \ x \ \frac{12}{5} = \frac{78}{5} \ or \ 15\frac{3}{5}$

4. $6 \ x \ 4\frac{5}{6} = \frac{6}{1} \ x \ \frac{29}{6} = \frac{29}{1} \ or \ 29$

5. $4\frac{2}{3} \ x \ 1\frac{5}{7} = \frac{14}{3} \ x \ \frac{12}{7} = \frac{8}{1} \ or \ 8$

6. $1\frac{7}{8} \ x \ 1\frac{3}{7} = \frac{15}{8} \ x \ \frac{10}{7} = \frac{75}{28} \ or \ 2\frac{19}{28}$

Mixed Practice

1. $3\frac{1}{3} + 1\frac{4}{5} =$

2. $7\frac{1}{2}\ x\ 1\frac{4}{5} =$

3. $6\frac{1}{2} - 1\frac{3}{8} =$

4. $2\frac{1}{6}\ x\ 1\frac{3}{5} =$

5. $2\frac{5}{7} - 1\frac{3}{4} =$

6. $5\frac{2}{3} + 1\frac{7}{8} =$

Check Your Answers

50 + 27

1. $3\frac{1}{3} + 1\frac{4}{5} = \frac{10}{3} + \frac{9}{5} = \frac{77}{15} \quad or\ 5\frac{2}{12}$

2. $7\frac{1}{2}\ x\ 1\frac{4}{5} = \frac{\overset{3}{\cancel{15}}}{2}\ x\ \frac{9}{\underset{1}{\cancel{5}}} = \frac{27}{2} \quad or\ 13\frac{1}{2}$

104-22

3. $6\frac{1}{2} - 1\frac{3}{8} = \frac{13}{2} - \frac{11}{8} = \frac{82}{16} = \frac{41}{8} \quad or\ 5\frac{1}{8}$

4. $2\frac{1}{6}\ x\ 1\frac{3}{5} = \frac{13}{\underset{3}{\cancel{6}}}\ x\ \frac{\overset{4}{\cancel{8}}}{5} = \frac{52}{15} \quad or\ 3\frac{7}{15}$

76 - 49

5. $2\frac{5}{7} - 1\frac{3}{4} = \frac{19}{7} - \frac{7}{4} = \frac{27}{28}$

136+45

6. $5\frac{2}{3} + 1\frac{7}{8} = \frac{17}{3} + \frac{15}{8} = \frac{181}{24} \quad or\ 7\frac{13}{24}$

***Remember the **FIRST STEP** with any mixed number or whole number is to change them into improper fractions.

Remember you can **ONLY** reduce across ***multiplication signs.

***Reduce when you can, but be careful not to over reduce. Make sure you reduce only at the right time for the operation you are doing.

***Remember that when doing adding and subtracting with mixed numbers that the numerators of the improper fraction can be quite large. Don't forget to multiply the opposite numerator and denominator before you add or subtract the new numerators.

Hold

Everything!!

The RULES are a CHANGING !

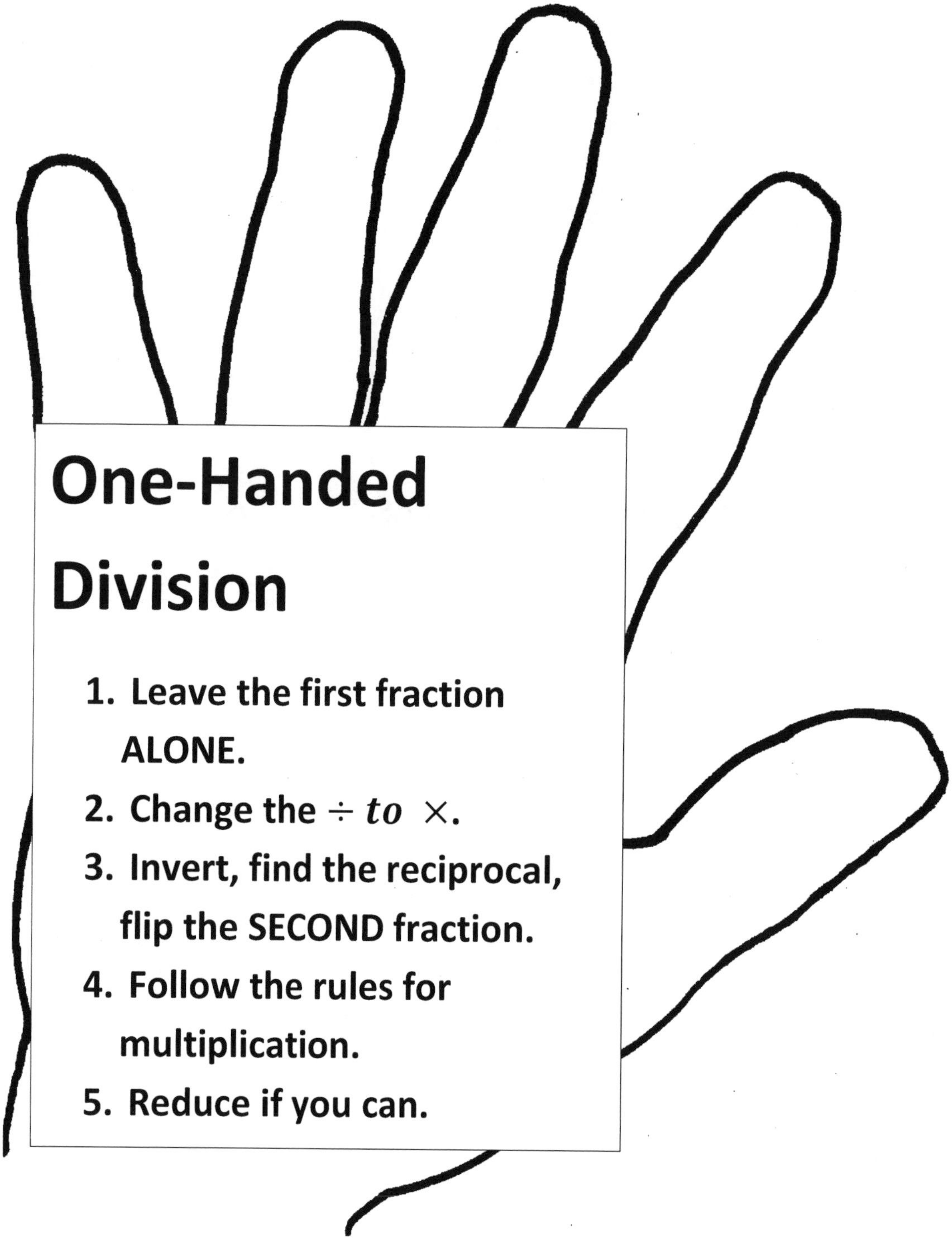

One-Handed Division

1. Leave the first fraction ALONE.
2. Change the $\div$ to $\times$.
3. Invert, find the reciprocal, flip the SECOND fraction.
4. Follow the rules for multiplication.
5. Reduce if you can.

Division of Fractions

***Add and Subtract fractions

1. Multiply.
2. Multiply.
3. Multiply.
4. Add or Subtract new numerators.
5. Reduce if you can.

***Multiplication of fractions

1. Reduce.
2. Reduce.
3. Multiply numerators.
4. Multiply denominators.
5. Reduce if you can.

Division has its own set of rules.

$$\frac{3}{4} \div \frac{5}{8} =$$

1. Leave the first fraction alone. $\frac{3}{4}$
2. Change the $\div$ to x. $\frac{3}{4}\ x$
3. Invert, find the reciprocal, flip the second fraction. $\frac{3}{4}\ x \frac{8}{5} =$

   You ALWAYS flip the second fraction no matter what it looks like. The first fraction is NEVER flipped. Remember that invert means to flip or do a 180.
4. Follow the multiplication rules. $\frac{3}{1}\ x \frac{2}{5} = \frac{6}{5}\quad or\ 1\frac{1}{5}$
5. Reduce if you can.

Examples

1. $$\frac{2}{3} \div \frac{20}{21} = \frac{\overset{1}{\cancel{2}}}{\underset{1}{\cancel{3}}}\ x\ \frac{\overset{7}{\cancel{21}}}{\underset{10}{\cancel{20}}} = \frac{7}{10}$$

2. $$\frac{15}{16} \div \frac{7}{24} = \frac{15}{\underset{2}{\cancel{16}}}\ x\ \frac{\overset{3}{\cancel{24}}}{7} = \frac{45}{14}\ \ or\ 3\frac{3}{14}$$

3. $$\frac{9}{10} \div \frac{27}{50} = \frac{\overset{1}{\cancel{9}}}{\underset{1}{\cancel{10}}}\ x\ \frac{\overset{5}{\cancel{50}}}{\underset{3}{\cancel{27}}} = \frac{5}{3}\ \ or\ \ 1\frac{2}{3}$$

4. $$\frac{16}{21} \div \frac{32}{35} = \frac{\overset{1}{\cancel{16}}}{\underset{3}{\cancel{21}}}\ x\ \frac{\overset{5}{\cancel{35}}}{\underset{2}{\cancel{32}}} = \frac{5}{6}$$

Problems for you to try

1. $\frac{3}{8} \div \frac{15}{16} =$

2. $\frac{16}{17} \div \frac{12}{25} =$

3. $\frac{28}{45} \div \frac{21}{40} =$

4. $\frac{5}{9} \div \frac{25}{36} =$

5. $\frac{24}{35} \div \frac{12}{21} =$

6. $\frac{4}{5} \div \frac{16}{25} =$

Check your work

1. $\frac{3}{8} \div \frac{15}{16} = \frac{\overset{1}{\cancel{3}}}{\underset{1}{\cancel{8}}}\ x\ \frac{\overset{2}{\cancel{16}}}{\underset{5}{\cancel{15}}} = \frac{2}{5}$

2. $\frac{16}{17} \div \frac{12}{25} = \frac{\overset{4}{\cancel{16}}}{17}\ x\ \frac{25}{\underset{3}{\cancel{12}}} = \frac{100}{51} \quad or\ 1\frac{49}{51}$

3. $\frac{28}{45} \div \frac{21}{40} = \frac{\overset{4}{\cancel{28}}}{\underset{9}{\cancel{45}}}\ x\ \frac{\overset{8}{\cancel{40}}}{\underset{3}{\cancel{21}}} = \frac{32}{27} \quad or\ 1\frac{5}{27}$

4. $\frac{5}{9} \div \frac{25}{36} = \frac{\overset{1}{\cancel{5}}}{\underset{1}{\cancel{9}}}\ x\ \frac{\overset{4}{\cancel{36}}}{\underset{5}{\cancel{25}}} = \frac{4}{5}$

5. $\frac{24}{35} \div \frac{12}{21} = \frac{\overset{2}{\cancel{24}}}{\underset{5}{\cancel{35}}}\ x\ \frac{\overset{3}{\cancel{21}}}{\underset{1}{\cancel{12}}} = \frac{6}{5} \quad or\ 1\frac{1}{5}$

6. $\frac{4}{5} \div \frac{16}{25} = \frac{\overset{1}{\cancel{4}}}{\underset{1}{\cancel{5}}}\ x\ \frac{\overset{5}{\cancel{25}}}{\underset{4}{\cancel{16}}} = \frac{5}{4} \quad or\ 1\frac{1}{4}$

***Remember do not try to reduce until you have the x sign even if the fractions will not reduce once you have flipped the second one.

***ALWAYS invert the second fraction no matter what it looks like. There are no exceptions to this rule.

Mixed Practice

1. $\frac{4}{7}+\frac{3}{8}=$

2. $\frac{7}{9}\,x\,\frac{3}{5}=$

3. $\frac{16}{25}\div\frac{12}{35}=$

4. $\frac{6}{7}-\frac{1}{3}=$

5. $\frac{6}{7}\,x\,\frac{35}{36}=$

6. $\frac{7}{11}\div\frac{22}{35}=$

7. $\frac{8}{9}+\frac{2}{3}=$

8. $\frac{7}{8}-\frac{3}{5}=$

Check Your Answers

32+21

1. $\frac{4}{7} + \frac{3}{8} = \frac{53}{56}$

2. $\frac{7}{\cancel{9}_{3}}\ x\ \frac{\cancel{3}^{1}}{5} = \frac{7}{15}$

3. $\frac{16}{25} \div \frac{12}{35} = \frac{{}^{4}\cancel{16}}{\cancel{25}_{5}}\ x\ \frac{\cancel{35}^{7}}{\cancel{12}_{3}} = \frac{28}{15}\quad or\ 1\frac{13}{15}$

18 - 7

4. $\frac{6}{7} - \frac{1}{3} = \frac{11}{21}$

5. $\frac{{}^{1}\cancel{6}}{\cancel{7}_{1}}\ x\ \frac{\cancel{35}^{5}}{\cancel{36}_{6}} = \frac{5}{6}$

6. $\frac{7}{11} \div \frac{22}{35} = \frac{7}{11}\ x\ \frac{35}{22} = \frac{105}{242}$

24 + 18

7. $\frac{8}{9} + \frac{2}{3} = \frac{42}{27} = \frac{14}{9}\quad or\ 1\frac{5}{9}$

35 - 24

8. $\frac{7}{8} - \frac{3}{5} = \frac{11}{40}$

***Division problems have an extra step over multiplication problems. Do not try to reduce across a division sigh even if it will not reduce once you have the x sign.

***Any time you get confused, go back and check the steps. Try not to practice a problem with the wrong steps.

***Don't worry if your answer looks weird as long as you have followed the steps and you have not made a calculation mistake, you are good to go.

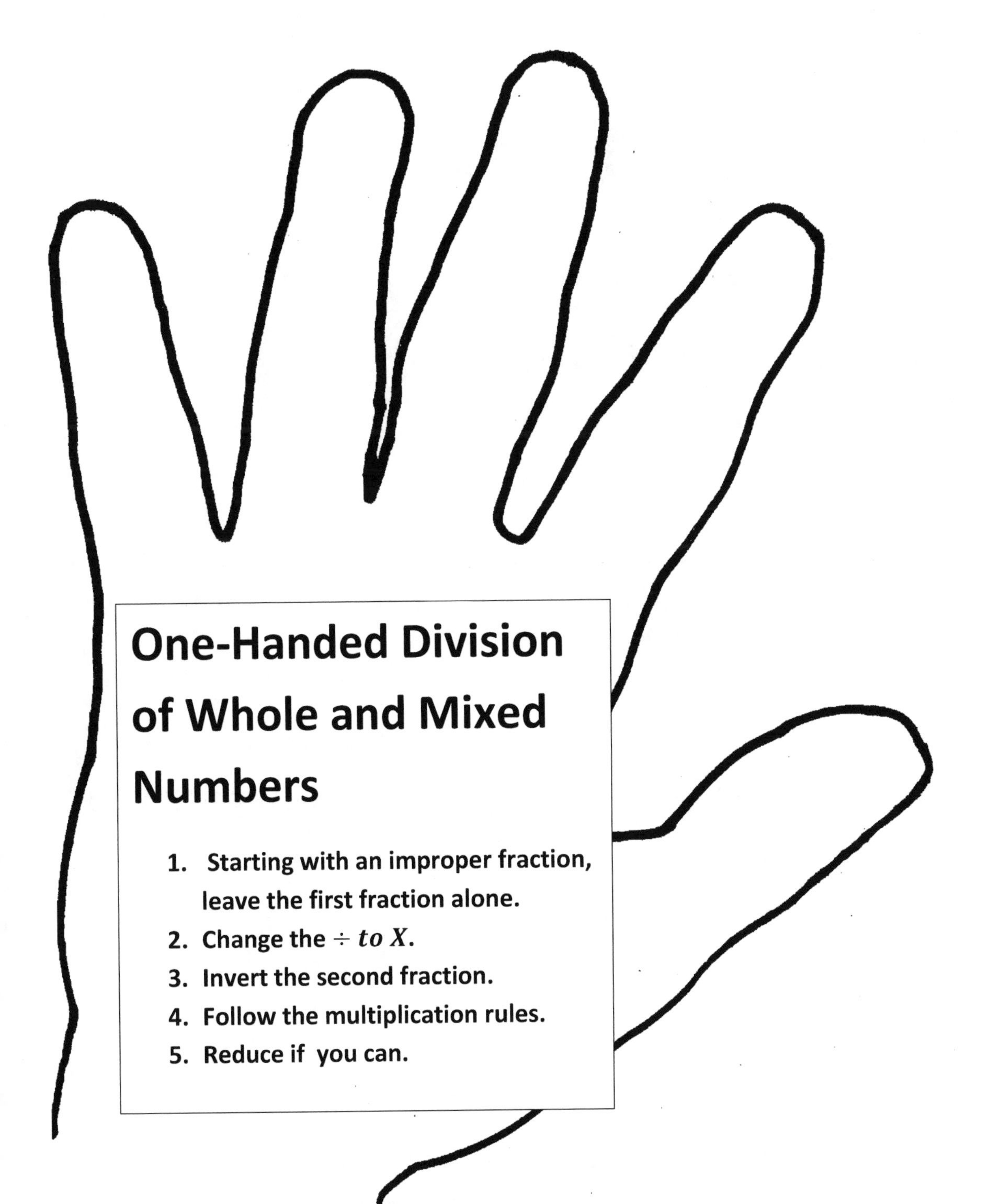
One-Handed Division of Whole and Mixed Numbers
1. Starting with an improper fraction, leave the first fraction alone.
2. Change the $\div$ *to X*.
3. Invert the second fraction.
4. Follow the multiplication rules.
5. Reduce if you can.

SKILL STOP

ALWAYS change a mixed number or whole number to an improper fraction before adding, subtracting, multiplying, or dividing.

- If you have forgotten how to change to improper fractions, look back on page 7.
- **After changing to an improper fraction follow the 5 step, One-Handed method for that operation.**

*****These problems have an extra step over and above all the other fractions we have done to this point. It is important not to try to change to improper fractions and invert the second fraction in one step. When you try to do this, you will forget to invert much of the time.**

Example

$$1\frac{1}{3} \div 1\frac{1}{9} = \frac{4}{3} \div \frac{10}{9} = \frac{\overset{2}{\cancel{4}}}{\underset{1}{\cancel{3}}} x \frac{\overset{3}{\cancel{9}}}{\underset{5}{\cancel{10}}} = \frac{6}{5} \quad or \quad 1\frac{1}{5}$$

Steps

1. Leave the first fraction alone
2. Change the $\div$ to x.
3. Invert, find the reciprocal, flip the second fraction.
4. Follow the rules for multiplication.
5. Reduce if you can.

*****As you can see, the rules for addition, subtraction, multiplication, and division of Mixed Numbers and Whole Numbers are not different from the rules for adding, subtracting, multiplying, and dividing fractions once you have changed to improper fractions.**

Examples

1. $4\frac{2}{3} \div 1\frac{7}{9} = \frac{14}{3} \div \frac{16}{9} = \frac{\overset{7}{\cancel{14}}}{\underset{1}{\cancel{3}}}\ x\ \frac{\overset{3}{\cancel{9}}}{\underset{8}{\cancel{16}}} = \frac{21}{8} \quad or\ 2\frac{5}{8}$

2. $3\frac{1}{3} \div 5 = \frac{10}{3} \div \frac{5}{1} = \frac{\overset{2}{\cancel{10}}}{3}\ x\ \frac{1}{\underset{1}{\cancel{5}}} = \frac{2}{3}$

3. $1\frac{2}{5} \div 1\frac{3}{7} = \frac{7}{5} \div \frac{10}{7} = \frac{7}{5}\ x\ \frac{7}{10} = \frac{49}{50}$

4. $6\frac{4}{9} \div 5\frac{1}{3} = \frac{58}{9} \div \frac{16}{3} = \frac{\overset{29}{\cancel{58}}}{\underset{3}{\cancel{9}}}\ x\ \frac{\overset{1}{\cancel{3}}}{\underset{8}{\cancel{16}}} = \frac{29}{24} \quad or\ 1\frac{5}{24}$

5. $2\frac{1}{4} \div 3\frac{5}{6} = \frac{9}{4} \div \frac{23}{6} = \frac{9}{\underset{2}{\cancel{4}}}\ x\ \frac{\overset{3}{\cancel{6}}}{23} = \frac{27}{46}$

Problems to Try

1. $5\frac{1}{2} \div 1\frac{1}{10} =$

2. $3\frac{3}{4} \div 4\frac{3}{8} =$

3. $2\frac{5}{6} \div 1\frac{1}{12} =$

4. $1\frac{3}{4} \div 12 =$

5. $6\frac{2}{3} \div 2\frac{2}{9} =$

6. $11\frac{1}{4} \div 3\frac{3}{8} =$

Check Your Answers

1. $5\frac{1}{2} \div 1\frac{1}{10} = \frac{11}{2} \div \frac{11}{10} = \frac{\cancel{11}^{1}}{\cancel{2}_{1}} \; x \; \frac{\cancel{10}^{5}}{\cancel{11}_{1}} = \frac{5}{1} \quad or \;\; 5$

2. $3\frac{3}{4} \div 4\frac{3}{8} = \frac{15}{4} \div \frac{35}{8} = \frac{\cancel{15}^{3}}{\cancel{4}_{1}} \; x \; \frac{\cancel{8}^{2}}{\cancel{35}_{7}} = \frac{6}{7}$

3. $2\frac{5}{6} \div 1\frac{1}{12} = \frac{17}{6} \div \frac{13}{12} = \frac{17}{\cancel{6}_{1}} \; x \; \frac{\cancel{12}^{2}}{13} = \frac{34}{13} \quad or \;\; 2\frac{8}{13}$

4. $1\frac{3}{4} \div 12 = \frac{7}{4} \div \frac{12}{1} = \frac{7}{4} \; x \; \frac{1}{12} = \frac{7}{48}$

5. $6\frac{2}{3} \div 2\frac{2}{9} = \frac{20}{3} \div \frac{20}{9} = \frac{\cancel{20}^{1}}{\cancel{3}_{1}} \; x \; \frac{\cancel{9}^{3}}{\cancel{20}_{1}} = \frac{3}{1} \quad or \;\; 3$

6. $11\frac{1}{4} \div 3\frac{3}{8} = \frac{45}{4} \div \frac{27}{8} = \frac{\cancel{45}^{5}}{\cancel{4}_{1}} \; x \; \frac{\cancel{8}^{2}}{\cancel{27}_{3}} = \frac{10}{3} \quad or \;\; 3\frac{1}{3}$

Try Some Mixed Practice

1. $2\frac{3}{4} + 3\frac{1}{3} =$

2. $5\frac{1}{3} - 2\frac{5}{6} =$

3. $5\frac{1}{4} \div 2\frac{3}{16} =$

4. $3\frac{2}{3}\ x\ 2\frac{2}{11} =$

5. $13 - 6\frac{4}{11} =$

6. $6\frac{1}{2}\ x\ 2\frac{2}{11} =$

7. $2\frac{3}{5} + 4\frac{1}{2} =$

8. $5\frac{1}{3} \div 2\frac{2}{9} =$

Check Your Answers

33 + 40

1. $2\frac{3}{4} + 3\frac{1}{3} = \frac{11}{4} + \frac{10}{3} = \frac{73}{12}$ or $6\frac{1}{12}$

96 - 51

2. $5\frac{1}{3} - 2\frac{5}{6} = \frac{16}{3} - \frac{17}{6} = \frac{45}{18} = \frac{15}{6} = \frac{5}{2}$ or $2\frac{1}{2}$

3. $5\frac{1}{4} \div 2\frac{3}{16} = \frac{21}{4} \div \frac{35}{16} = \frac{\overset{3}{\cancel{21}}}{\underset{1}{\cancel{4}}} \; x \; \frac{\overset{4}{\cancel{16}}}{\underset{5}{\cancel{35}}} = \frac{12}{5}$ or $2\frac{2}{5}$

4. $3\frac{2}{3} \; x \; 2\frac{2}{11} = \frac{\overset{1}{\cancel{11}}}{\underset{1}{\cancel{3}}} \; x \; \frac{\overset{8}{\cancel{24}}}{\underset{1}{\cancel{11}}} = \frac{8}{1}$ or 8

143-70

5. $13 - 6\frac{4}{11} = \frac{13}{1} - \frac{70}{11} = \frac{73}{11}$ or $6\frac{7}{11}$

6. $6\frac{1}{2} \; x \; 4\frac{1}{2} = \frac{13}{2} \; x \; \frac{9}{2} = \frac{117}{4}$ or $29\frac{1}{4}$

26+45

7. $2\frac{3}{5} + 4\frac{1}{2} = \frac{13}{5} + \frac{9}{2} = \frac{71}{10}$ or $7\frac{1}{10}$

8. $5\frac{1}{3} \div 2\frac{2}{9} = \frac{16}{3} \div \frac{20}{9} = \frac{\overset{4}{\cancel{16}}}{\underset{1}{\cancel{3}}} \; x \; \frac{\overset{3}{\cancel{9}}}{\underset{5}{\cancel{20}}} = \frac{12}{5}$ or $2\frac{2}{5}$

Reduce all final answers to lowest terms. You can leave an improper fraction as your answer as long as it is reduced.

- Do not over reduce, not all fractions can be reduced.
- Most of the time in higher math, you leave the improper fraction as the answer. But there will be times when you will need to change to a mixed #.
- Be careful of fractions like $\frac{16}{21}$. The numbers both have factors, but not any common ones, so the fraction cannot be reduced.
- Be careful with 6. Many mistakes are made here because you get going too fast and forget whether you are dividing by 2 or 3.

➢ Don't flip any fractions with multiplication, addition, or subtraction and never flip the first fraction in division.
➢ Don't reduce across any sign except multiplication, X.

Add or Subtract Fractions

- ❖ Multiply.
- ❖ Multiply.
- ❖ Multiply.
- ❖ Add or subtract numerator.
- ❖ Reduce if can.

Multiply Fractions

- ❖ Reduce numerator and opposite denominator if can.
- ❖ Reduce other opposite numerator and denominator if can.
- ❖ Multiply numerators.
- ❖ Multiply denominators.
- ❖ Reduce if can.

Divide Fractions

- ❖ **ALWAYS** leave first fraction **ALONE.**
- ❖ Change the ÷ to X.
- ❖ **ALWAYS** flip, invert, find the reciprocal of the **SECOND** fraction.
- ❖ Follow the multiplication rules.
- ❖ Reduce if can.

*****Always change a Mixed Number or Whole Number to an improper fraction and then follow the above steps.**

OK, no more rules or rule changes! I truly hope this book has helped you with the operations of fractions. Fractions are truly not hard as long as you remember the 5 steps that go with each operation.

I would love to hear from you once you have finished the book, or if you have problems before you finish. Please do not give up on fractions if you hit a snag that does not make sense to you. Find some help.

The last two pages are meant to be pulled out and used as reference once you have the process down well enough to only need a small reminder every once and a while. Remember you will see fractions in all types of math classes! You can do them, no need to skip them!

Please email me, karenmath58@yahoo.com. I look forward to hearing from you.

Karen Tollefson